Woodlot Management

by Jay Heinrichs

As the demand for renewable resources such as trees escalates, and as rules for taking wood out of government land get stricter, the forest industry predicts that it is going to turn more to the small-woodlot owner for forest products. That will mean higher prices for wood, and more intensive lobbying for woodlot tax relief.

If you own less than ten acres of forest, you may not find it practical to manage your woods for timber production. But you can produce healthier, faster-growing trees and supply more wood for your stove. In addition, a few exceptionally valuable old hardwood trees might be worth a logger's trouble to remove them individually — and could bring you a tidy sum.

Most woodlots are overcrowded, with competition among the trees so intense that the wood grows about half as fast as it could. Without careful management, opportunities for recreation are diminished, and the forest does not support the wildlife that it could.

There are people who say you should walk softly on the land and take care not to disturb the natural balance of the forest. These same people will tell you to cut only what you need from your forests and let nature take care of itself. Well, if humans had never walked the earth, this system of forest management would work just fine. But in most wooded tracts in the United States, the natural balance has been seriously upset, and a great deal of work is necessary to set it right.

There is a real temptation to avoid the long wait for trees to grow valuable. But using your woodlot just for burning or selling firewood for a fast buck can be like burning your dining room furniture to cook dinner. It pays to plan ahead. That is what this bulletin is for: to offer advice on how to have your woodland home and heat it, too. We will outline sound forestry practices for managing your woodlands for fuel wood, timber, maple sugaring, or recreational uses. Finally, we will point you in the direction of some additional help available to you, the owner of a woodlot.

How Much Forest Do You Need?

Even with a relatively small woodlot, you can produce enough wood to heat your home forever. Exactly how small it can be depends on the energy efficiency and size of your home, and the efficiency of your stove. In the old days, when wood was there for the taking, inefficient stoves and fireplaces burned wood in large, drafty houses at the rate of ten to fifteen cords per year. But the average modern house in a northern climate can now be heated year-round with only three to eight cords.

Growth Per Acre

The growth rate per acre in most forests is between one-quarter and three-quarters of a cord per year, depending in part on the climate, soil, kinds of trees, and the degree of management. In much of the East, for example, the volume of wood in the forests doubles every ten to twenty years. If it is harvested under intensive-management techniques, a good crop of trees can be obtained on the same woodlot every twelve to fifteen years. Most foresters say that a vast majority of woodlots in this country could double the amount of wood they produce — if cared for properly. Some of the management techniques include thinning, selective cutting, planting faster-growing species of trees, and harvesting trees before they become "overmature" and slow their growth rate. Many trees send out shoots from stumps, which develop into fast-growing trees whose root systems are already established. Usually, these sprouted trees do not grow straight and tall for timber production, but they make perfectly good firewood.

In short, a carefully managed woodlot need be only five acres or more to give you year-round fuel for heating and cooking. If your woodlot is larger, you might have enough wood left over to sell some firewood. If it is larger than ten acres, you should get professional or governmental help and consider managing it for timber. But before you do anything to your woods, you should go see what you have.

Woodland Inventory

Part of a good forest inventory involves making sure the trees you cut are yours. So, if you have not done so already, get your neighbors to walk along the borders of your lot with you. In many woodlots, boundary corners are marked with metal rods or small piles of rocks. Mark the boundaries themselves by painting boundary rocks. You can also paint boundary trees; but if you do, use quick-drying enamel or a caulking compound. Aerosol paint can lead to decay in trees. Use aerosol spray only to mark trees that are about to be cut. Be especially careful to mark the points where the boundary lines change direction.

The Soil Conservation Service in the U.S. Department of Agriculture can supply you with a map or aerial photos that can give you a general idea of the shape of your woodland, and should tell something about the kinds of trees on it. Soil maps can tell you how good the land is for growing trees. The agricultural extension agent for your county should be listed in the phone book; he or she can tell you how to get maps and photos.

Know Your Trees

Now for inventory itself. The most important expertise you need to carry into the woods with you is the ability to distinguish between different species of trees. A good tree guide is indispensable. You cannot beat the *Audubon Society Field Guide to North American Trees* by Elbert L. Little. It comes in both eastern and western region editions.

The best wood for burning is *hardwood*, found in most broadleaf (deciduous) trees. Most of these trees lose their leaves in the winter. *Softwood*, found in needleleaf (evergreen) and some deciduous trees, can also be burned, although the burning qualities of softwood are inferior to those of hardwood. But in some parts of the country, you have no choice — you may not *have* any hardwood.

Although you will do most of your firewood cutting in the late fall and early winter, it is a good idea to take inventory in the spring or summer. That is when the leaves and fruit are on most of the trees, making them easy to identify. You can distinguish a tree

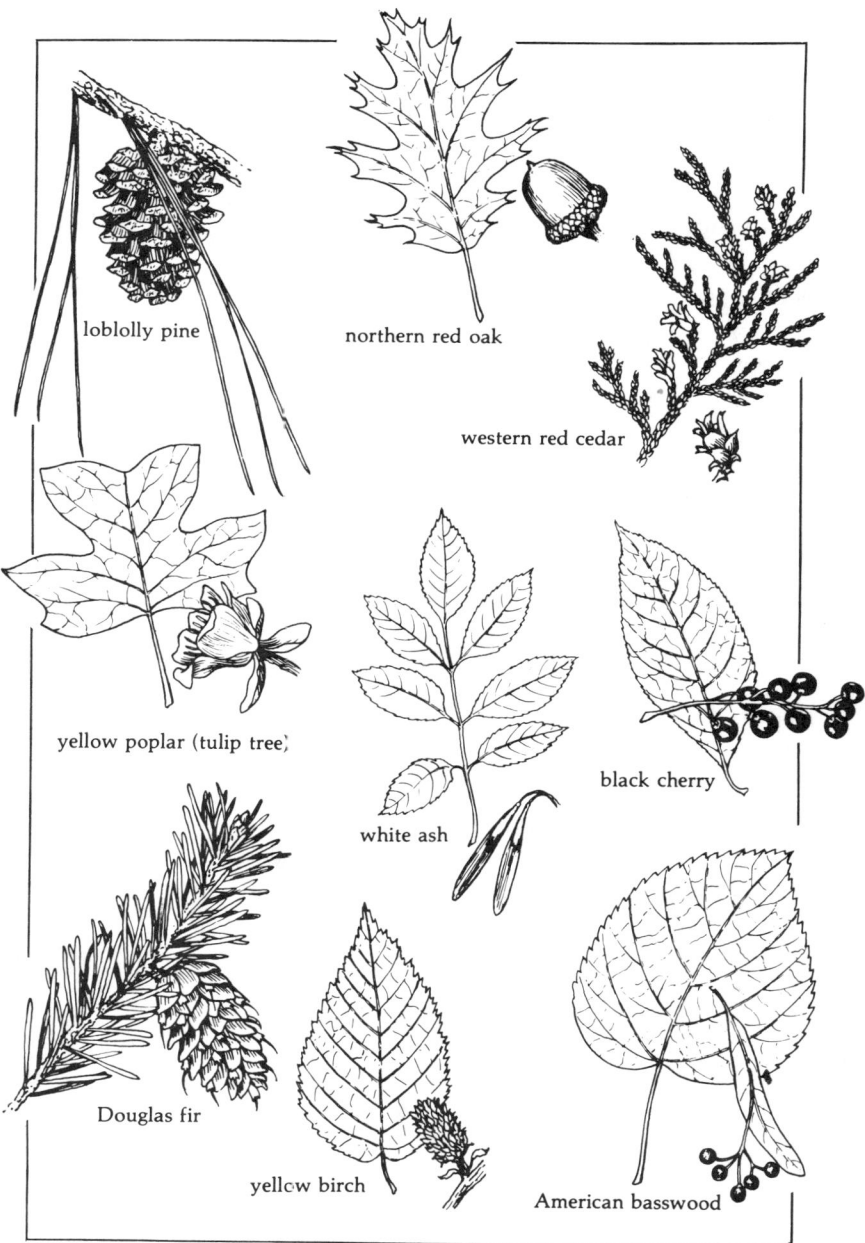

In the spring and summer, you can identify trees by their seeds and leaves. Here is a sampling of some good timber trees.

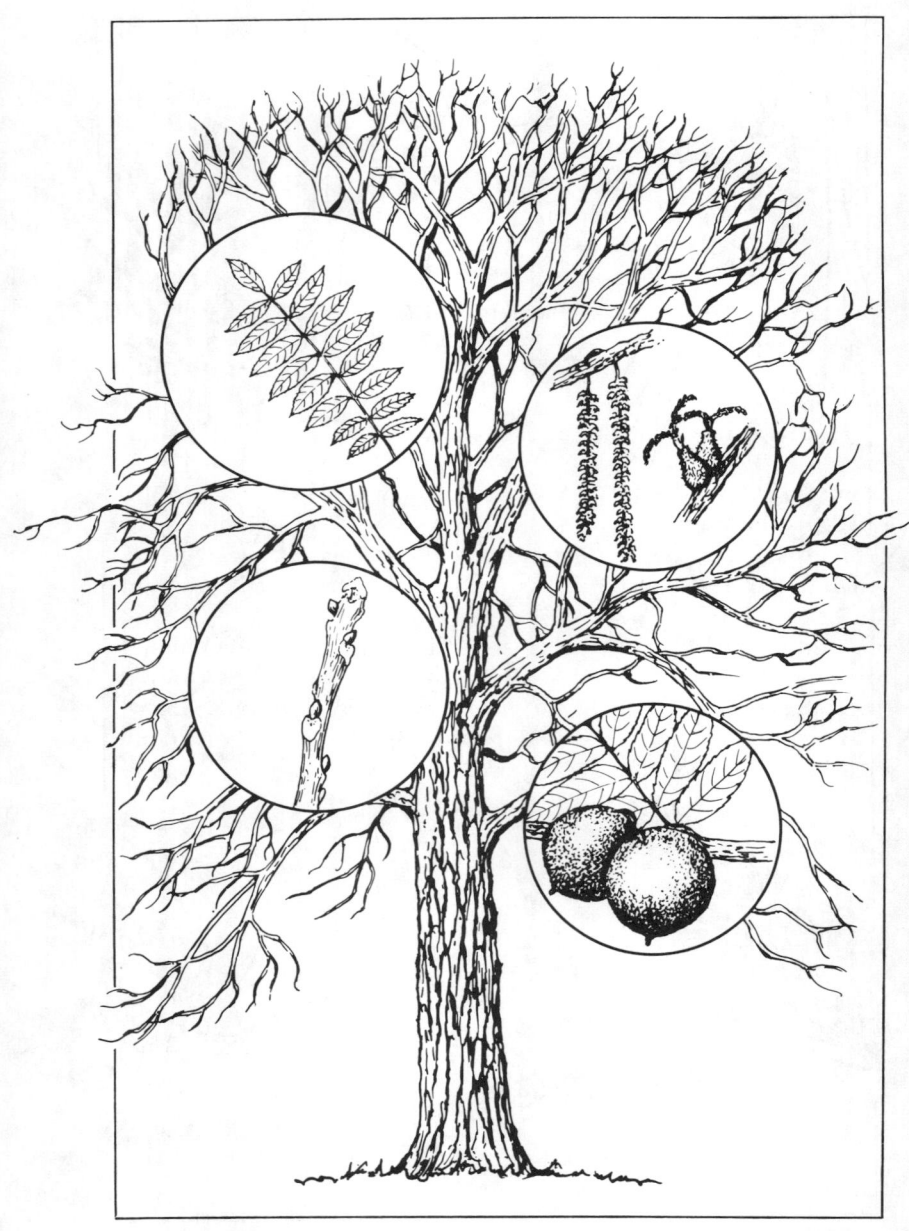

Here a black walnut is identified. Notice the scaly bark, the compound leaves, the alternate buds on the twig, the characteristic fruit and flowers.

by the color, texture, and the smell of its bark; by the characteristics of its leaves, fruit, and twigs; and by the shape and size of the whole tree.

It is very important to be able to identify your trees. If you fail to recognize an enormous black walnut and burn it for firewood, for example, you could put yourself out of several thousand dollars. If you are considering managing your woods for timber or pulp production, choice species include white ash, basswood, yellow birch, black cherry, cucumber tree, the southern cypresses, Douglas fir, white and northern oak, white and red pines, many of the southern pines, sugar maple, sweet gum, and yellow poplar. In the Pacific Northwest, western red cedar is prized for making shingles and shakes.

Ralph Waldo Emerson said that a weed is "a plant whose virtues have not yet been discovered." Many trees that are not valuable today may bring good prices a few decades from now, so it helps to be cagey and to consult a good forester. But some species probably will never sell well among the timber and pulp industries. Species that do not have much of a market as timber, but make excellent firewood, include beech, sweet birch, box elder, the buckeyes, butternut, cottonwood, the elms, hackberry, tupelo, and others. If you don't mind taking down some beautiful trees, you can cut quaking aspen, grey birch, pin cherry, flowering dogwood, American holly, ironwood (eastern hop hornbeam), American hornbeam, and rosebay rhododendron for firewood.

Managing for Firewood

The key to managing for firewood is knowing what to cut and when. A *selective cut,* in which you carefully choose the trees you want to take out, is the opposite of a *clearcut,* in which all the trees are harvested indiscriminately. Clearcuts result in *even-age management,* because when the stand is reforested, all the trees that grow back are the same age. A selective cut allows you to ensure a variety of ages in your forest. Although even-age management is good in many situations, in most cases your woodlot is better off with a mixture of ages. A mixed-age forest is less susceptible to diseases and pests, and it is more interesting to look at.

Selective Cutting

Plan to remove the poorly growing wolf trees, and the trees that are growing skinny and crooked. In addition, you may have to remove some of the best-growing trees — the "dominant crown trees." These are the ones whose branches reach over the branches of others, and which have the most opportunity for growth. These dominant crown trees should be carefully spaced for maximum growth.

Cutting Formula. To determine whether two of these dominant crown trees are too close to each other, take their average diameter in inches (add them and divide by two) and then add six (a constant figure). The resulting figure is the proper distance *in feet* between the dominant trees. For example, the distance between a twelve-inch dominant tree and a twenty-two-inch dominant tree should be twenty-three feet (12 + 22 divided by 2, plus 6, equals 23). But use your judgment; some trees with particularly large crowns need more space. Others need less. Hardwoods generally need more room in the forest than softwoods. If you are managing your woodlot for firewood, and you find two straight, tall trees right next to each other, cut down the larger one. But if you are managing your land for saw timber, cut the smaller one, leaving the crop tree for future harvesting.

When you determine which dominant tree to cut, do not take out

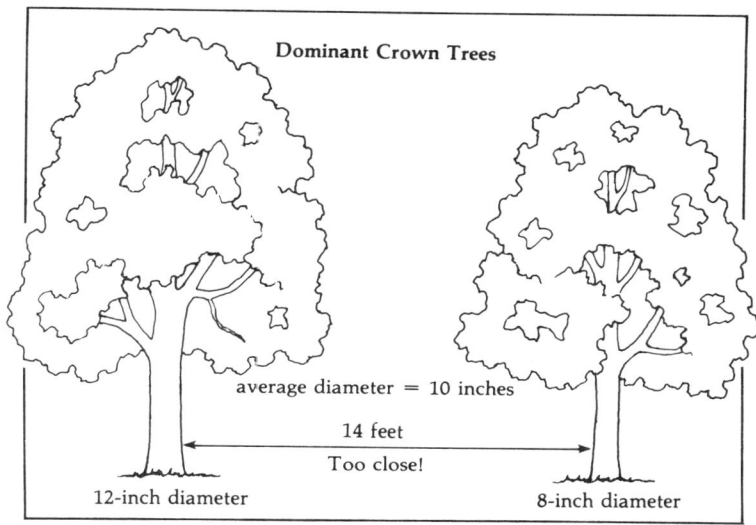

To determine whether two dominant crown trees are spaced too closely, take their average trunk diameter in inches and add 6. This gives you the proper spacing in feet. If you are managing for firewood, remove the larger tree. If you are managing for timber, you cull the smaller one.

everything between those trees; some smaller trees should be left. Plan to cut about the same number of small trees as large trees, leaving enough space for light and nutrients.

This does not mean necessarily that you should immediately go in and cut all that wood. If you cut too many trees at once, you will expose the forest soil to drying or invasion by shrubs, and you may let the wind in, which also dries the soil. Some competition among trees is good for *self-pruning*, in which a tree drops branches that have been shaded out. This produces a straight, tall tree that is relatively free of knots.

Clearcutting

If a portion of your woods has been badly abused and is growing a tangle of small, crooked trees, you might consider a clearcut. Despite all the bad things you have heard about logging abuses, a small clearcut can be sound forestry. It gives you a chance to plant the kind of trees you want to grow. Scientists have developed special fuel wood trees that grow very rapidly. The new hybrid

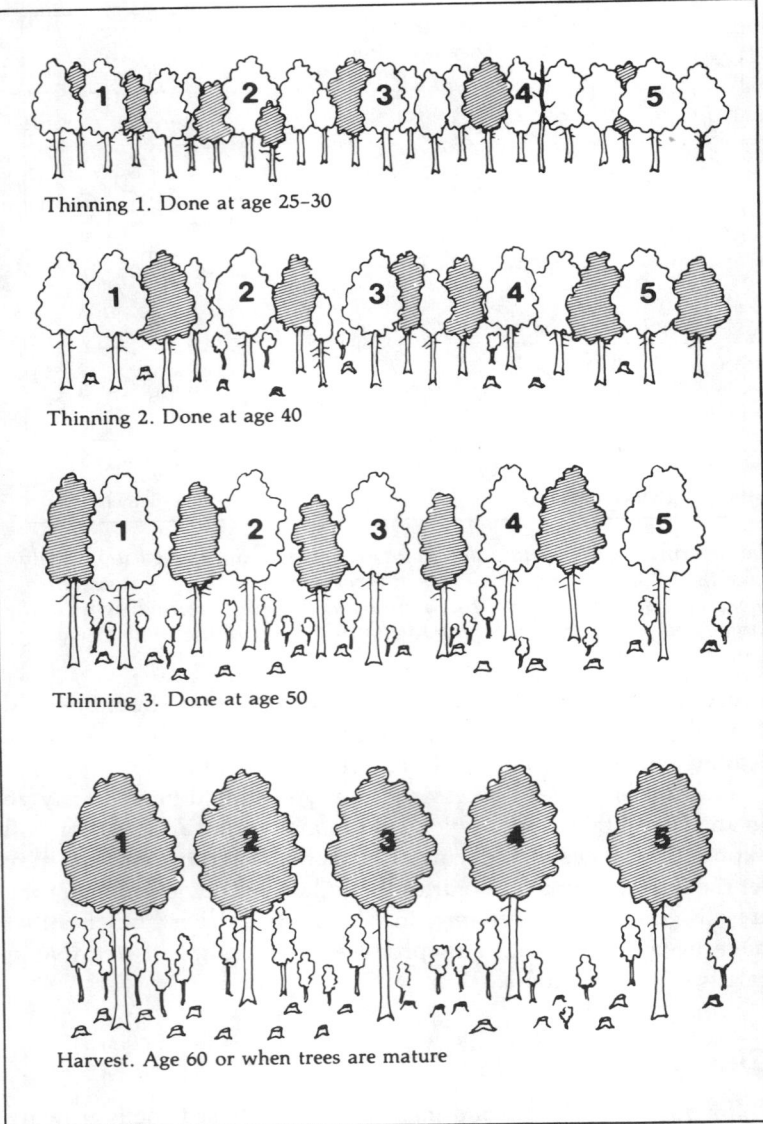

Thinning 1. Done at age 25–30

Thinning 2. Done at age 40

Thinning 3. Done at age 50

Harvest. Age 60 or when trees are mature

If your woodland was clearcut at one point, your timber stand will be all one age. Here is a thinning plan adapted from a publication of the Pennsylvania Department of Forests and Waters, Forest Plantation Management. At each thinning, cut the trees with the darkened crowns so that the numbered trees can mature without crowding.

poplars, for example, can grow to a foot in diameter in five to eight years. If used to stock a woodlot in the Northeast, they can almost double its energy production. Some states offer hybrid tree seedlings at nominal cost for firewood production. To learn more about clearcutting and reforestation, consult your county extension agent or state forester.

Tools for the Woodlot Owner

Now that you have decided what to cut and what not to cut, you have to prepare yourself for taking the trees down and processing them for burning. There are three basic steps to this operation: felling, bucking (cutting trees into four-foot or smaller logs), and splitting. You will need a number of fairly specialized tools; but you do not need all those that I list. Before you invest in hundreds or thousands of dollars worth of equipment, decide how much wood you will be cutting each year. It might be cheaper and easier to rent the more expensive equipment. Here are the tools you might want.

- chain saw
- crosscut saw
- felling wedges (plastic, aluminum, or magnesium)
- axes
- bow saw
- peavey
- splitting maul and steel wedges

If you are cutting a lot of firewood, you might want to invest in a mechanical wood splitter. There are all kinds of fancy wood splitters, but the best is the hydraulic splitter, which rams logs at 15,000 pounds per square inch. It costs over $1,500 to buy, but you can probably rent one to split several cords over a couple of days.

You can build your own hydraulic splitter operated by a three-horsepower gas engine. See the October 1976 issue of *Popular Mechanics* for details. Libraries often carry magazines from several years back. Or write directly to *Popular Mechanics* magazine.

Clothes

Proper clothing can increase your safety when you are using a chain saw. All your clothes should fit snugly, and by all means, tuck in your scarf before turning on your saw! You should also have goggles or glasses and a hardhat. Wearing a hardhat is always a good idea in the woods, especially during a period of high winds or drought.

The noise a chain saw makes is well above established safety levels. Buy shooters' earplugs or the muffs worn by airport-runway workers. You should wear heavy shoes for all work in the woods. Get loggers' knee pads — half of all chain saw accidents involve the knees.

Felling a Tree

The healthy respect you should have for your chain saw should extend to the tree the saw is cutting. Safety is one reason you should cut in the late fall: with the leaves off, you can easily determine which way a tree is leaning, and plan your attack accordingly. Another reason: wood dries better when cut in the fall. If you let a downed tree lie for a month or so, the bed of leaves can help absorb moisture. If allowed to sit through the summer, cut wood actually begins to reabsorb moisture. You can continue cutting into the winter — many hardwood stumps resprout better if they are cut when dormant. Otherwise the young sprouts of summer or early fall cannot stand the cold a few months later. One last reason for fall or winter cutting is convenience: the insects often leave the bark of a tree in early fall, and are less apt to invade your living room.

Before you actually start to cut, take a pair of lopping shears or a handsaw and clear away brush around the tree. Plan the landing zone — you should fell your tree in the general direction of its crown lean and the angle of its trunk. But you can "aim" the tree so that it does not get hung up in another tree or smash young saplings. Finally, plan your escape route and walk along it. When it comes time to clear out, you do not want any surprises tripping you up.

Notch cut. Your first cut is the notch cut, made on the side of the tree toward which it will fall. First, make a cut parallel to the ground more than a quarter and less than a third of the way into the trunk. Then cut down into this first cut to make a forty-five-

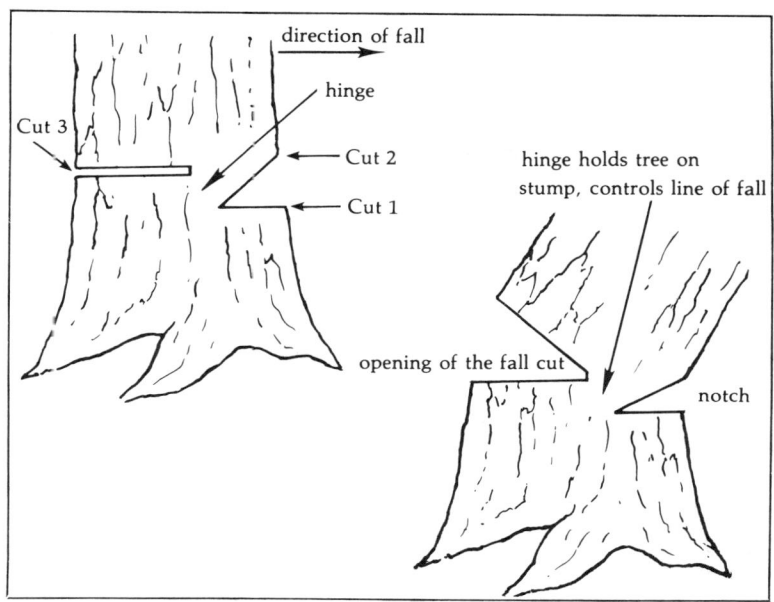

Felling a Tree

degree notch. Be careful when cutting into a wolf tree with several trunks; the trunks have a tendency to separate just when you finish the notch.

Fall cut. Now comes the back, or fall, cut, which should be parallel to the bottom of the notch cut and about three inches above it. The point is to create a hinge that controls the fall of the tree and keeps it from flipping up or rolling during descent. You can control the angle of fall slightly by making one side of the hinge fatter than the other; this is called "holding a corner." The tree will fall in the direction of the deepest cut.

Take it easy during the beginning of your fall cut, but keep your saw running at full speed. When you reach the middle of the cut, apply a little more pressure and cut as fast as the saw can comfortably go. This makes the tree go down faster and prevents hangups. There are several ways to release a hung-up tree. We recommend enlisting the help of an experienced friend. Such booby traps are best left to the experts.

If your saw gets bound while you are felling, stop the saw and tap a wedge or two (or three) into the back cut, behind the saw. The tree will tend to fall in the direction in which you point the wedge, so aim carefully. Use an ax or hammer and continue to tap the wedge in occasionally as you resume cutting.

Felling Wolf Trees

Wolf trees are tricky, even after you have made the notch cut. How you attack them depends on their shape, size, and the direction of their trunks. If a wolf tree is forked close to the ground, cut one trunk at a time and allow it to fall in the direction of its own lean. Cut a couple of feet above the fork to avoid the separation of another trunk. If the fork is too high for you to get your saw above it, you will have to cut below the fork. Remember that each trunk may act as a separate tree; but be aware that the other trunks may be affected by what you do to one. You may not be able to make a good back cut. Your back cut may actually have to be a side cut, supported by strategically placed wedges to help control the angle of fall. If the trunks are not too big, you can actually help push each one over. Try to remove the smallest trunks first, then work on the bigger ones.

Bucking and Splitting

Once the tree is down, the limbs should be removed (limbing) and the log cut into smaller pieces (bucking). First make sure the tree has settled properly on the ground. Then lop off trunk limbs, keeping the crown intact at first. When cutting large limbs, make your first cut from the bottom, then finish off from the top.

Buck the wood into lengths about four inches shorter than the length of your firebox. If you plan to sell the firewood, cut it into four-foot lengths. Cut all limbs three inches or more in diameter. Cut from the butt of the tree toward the crown, which acts as an anchor. The easiest way to buck is to cut downward until the saw threatens to bind. Then roll the log over with a peavey and finish the job. On large trees, saw about a fourth of the way through on the top of the log, then finish the cut from underneath. Keep the log off the ground, if you possibly can, to protect your chain saw.

Do not bother to cut branches less than three inches thick. Spread the leftover branches on the forest floor to decompose. Do not pile branches around the base of a tree — it's a fire hazard. But small piles of brush create good wildlife habitats.

Splitting

Although most moisture leaves logs through the ends, wood dries faster when it is split. Logs larger than eight inches in diameter should be split in quarters or smaller pieces.

Some woods split easier than others. Oak, ash, and beech should not cause you too much trouble. Hickory, elm, and ironwood are especially tough. Most wood splits easier when green; so split your wood shortly after cutting.

To split, use a splitting maul, and aim toward the center of the upended log. Split with the grain, taking advantage of cracks that exist naturally in the wood.

Drying

If you can possibly manage it, dry your wood in the woods, close to where it was cut. Dry wood is much lighter than green wood,

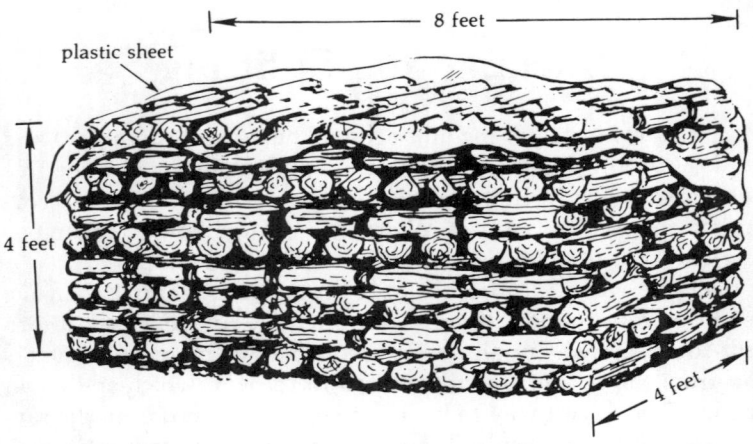

Wood dries faster when covered with clear plastic on top. The sides should be left free for air circulation. Shown here is a standard cord.

and so it is easier to transport. Wood should be dried to 20 percent moisture before it is suitable for burning. This takes six months to a year. Dry wood feels dry to the touch when the bark is peeled off, and its ends contain deep cracks. Check your wood carefully before you burn it; wood that is still green causes creosote to build up in your chimney, which can cause disastrous chimney fires.

Wood dries faster with a "solar drier," which is simply a sheet of heavy-duty clear plastic secured over your pile. Leave the end that faces the prevailing wind open for plenty of air circulation. Set logs lengthwise along the bottom of the pile for a base. Logs dry most effectively when stacked log-cabin style, with each level set at right angles to the one below it. If you do not cover your wood, stack it with the bark facing up. If you do cover it, the split sides should face up.

Selling Firewood

If you have more than enough wood and woodlot to grow your own fuel wood, you might consider selling some. The rocketing prices for firewood may give you a chance to cull much of your unwanted wood at once, getting your forest into shape very quickly.

Don't Sell Timber for Firewood

Call your county extension agent for free advice — and possible financial aid — for "timber-stand improvement." The advice could come in handy, because the first thing you should check when examining the wood market is whether you are cheating yourself by selling high-value wood for firewood. No matter how good the prices are for fuel, they will never beat the prices for lumber. And you might be surprised to learn what kind of wood is bought for other uses.

Call local pulp mills and sawmills to ask about the species bought, and prices paid, for delivered wood. The cooperative extension service offices in most states have reporting services that give prices for fuel wood, pulpwood, saw timber, peeler and veneer logs, chips, poles, crossties, fuel residues, and dimension lumber. Call your county extension agent or state forestry representative for information on this service.

Markets for Firewood

You can sell your firewood through local retailers and brokers. Again, make sure you know the value of your wood first. If possible, get yourself listed in your state's directory of fuel wood suppliers. Not all states have one, but an increasing number do. The simplest way to sell is to advertise the availability of "cut-your-own" stumpage, and let people come and remove it themselves. Mark the trees you want to come out. You should be able to build up a steady clientele over several seasons. But of course you cannot expect a lot of money per cord; one-fifth the price of delivered, bucked wood would be excellent money if you can get it. Ask the visitors to clean up after themselves. If you are unable or unwilling

> ## WHAT IS A CORD?
>
> A cord of wood is a stack 8 feet long, 4 feet deep, and 4 feet high. Its volume is therefore 128 cubic feet. But because the stack is not solid wood, about 20 percent of that volume is air. Different kinds of wood weigh differently, of course, but the average cord weighs about 1½ tons. It contains about thirty-five 10-inch-diameter logs 16 inches long, or about ten trees measuring 8 inches in diameter at about the height of your breast. (If you count the branches, though, a medium-size tree can produce half a cord.) An average cord produces the same heat energy as 166 gallons of No. 2 oil, or ¾ ton of hard coal, or about 5,000 kilowatt-hours of electricity.

to cut wood for your own use, you can let friends cut cord-wood "shares," and take as payment for yourself something like a third of each load. Before you advertise that you have cut-your-own wood, check the local liability laws; you may be legally responsible if someone is injured on your property.

More and more states are requiring firewood sellers to specify the volumes of their loads. An increasing number of jurisdictions forbid sales of "loads" or "racks" of wood, because the meaning of these terms is indefinite. Another unit is the "thrown" cord, which is wood thrown into a 144-cubic-foot truck body. Maine accepts this as a standard measure, but New Hampshire still uses the 128-cubic-foot cord of tightly stacked wood as a standard. Call your state forestry representative to see what regulations apply to you. Many departments establish fuel wood grades, specify permissible species and sizes, and state forms, condition of wood, sizes to be split, and so on. The state forester can also give you a good idea of fair firewood prices.

Some states allow firewood dealers to sell by the *face cord*, which is a stack of wood four feet high, eight feet long, and just about any depth four feet and under. A face cord is a useful measurement for selling wood that has been cut to a specified length to fit a firebox, but it is often used to cheat unwary buyers. When you sell a face cord, let your buyer know how much smaller it is than a standard cord.

Stack your firewood reasonably tight, but expect to be paid for a high-volume cord. The actual volume of wood in a standard cord varies between 60 and 100 cubic feet, depending on the kind of wood, the lengths and thicknesses of the logs, and how they are stacked. The cord with the least wood volume is the one in which the wood has been stacked one log on top of the other. The most wood volume is obtained when the logs are staggered, with small pieces squeezed between the gaps, and when the logs are short. If you sell your firewood by the truckload, make sure you know the volume of what you are selling. A cord of log lengths should sell for less than a cut-up cord.

One final piece of advice about delivering wood: use a good truck, and do not overload it. A cord of hardwood can weigh well over two tons. If you plan to sell wood regularly, do not use a car to deliver it! No load of firewood is worth the sacrifice of an axle. Drive slowly, and be careful when stopping — your brakes just won't seem as efficient when your truck is groaning under a full load.

Man Does Not Live On Fuel Alone

There are many uses of a forest that may be much more valuable than producing firewood. The most important of these is lumber. The U.S. Forest Service predicts that lumber demand will double by the year 2000 — a time when many of your more promising young trees could be large enough to sell at the promised higher prices. Call local sawmills and find out what kinds of wood are sold for lumber in your area, and what the prices are. Thanks to a rising housing demand, timber prices are rising twice as fast as the prices of other crops in this country.

But suppose your woodlot is a small one — ten acres, say. Is it worth a logger's while to take out such small amounts of timber? Is it worth your while to go through the trouble of selling the timber? All that depends on what you have on that small lot. A single huge, flawless black walnut tree can sell for several thousand dollars (one such tree reportedly brought $30,000 on the stump).

Before you consider selling any of your trees, get help from the government. A county or state forester will give you advice on whether you have timber to sell, and how to get more help to sell it. A word to the wise: *seek advice.* If you do not have experience in selling wood products, you risk ruining your woods and depleting your bank account. (See Help! in the back of this bulletin.)

One way to find out whether or not you have timber to sell is to conduct your own informal "cruise," or inventory, and see what you have. By conducting your own cruise, you can estimate the number of trees that should be removed to increase the vigor of the trees that remain. Then you can get a forester to help you find out the prices you can get for the trees you remove, and whether the trees are good for firewood or for other uses. If you are interested only in producing firewood, do not discount the possibility that you will want to sell timber in the distant future.

Examine a Sample Plot

The best way to estimate how many trees can be removed to increase the health of the woods is to examine a sample plot. First, mark the trees that you do *not* want to cut, for reasons of economic value, beauty, contribution to wildlife, or ability to restock the stand. Put a spot of paint at the base or tie a ribbon around each trunk. Then pick out a typical part of your woodland — if there is any typical part — a tenth of an acre in size. A tenth of an acre has a thirty-seven-foot radius, and you can measure a circular sample plot with a piece of rope that long. Within the sample plot you have measured out, pick out the ten best trees to save, counting the ones you have already marked. If the stand is composed of good softwoods, mark about twenty trees. The trees you mark should be the straightest, healthiest-looking trees. The leaves should be relatively clear of defects that might show disease. Where branches have naturally fallen off, the wounds should have healed cleanly.

Now count the number of trees that you want to cut down — the ones that you have not marked. That will determine how much firewood you have to use or sell. If you think that the "junk" wood might be good for something else, ask the government forester. Multiply the amount of "junk" by ten, and you will have the approximate number of firewood trees per acre. Take a look at the trees you have marked to save. If you can, get the forester to look at them. Do they show promise for timber? Did you have to look hard just to come up with ten good trees in your sample plot? Then you may need to get advice on a different form of management.

Small, single-acre clearcuts replanted with fast-growing trees might be a solution, or the forester might tell you how to manage for pulpwood production. Beware of choosing favored trees that are too close to each other; follow the formula mentioned on page 8, for thinning dominant crown trees.

The above inventory should be used only to help you decide what to do with your woodlands. If it appears you have a large number of big, slow-growing, healthy trees, you may already have a woodland sufficient for immediate timber sale. If so, get your county extension agent to recommend some good consulting foresters to help you with the sale.

The majority of small woodlots in this country, however, are in serious need of restoration. Most of them have suffered under a practice called *high-grading,* in which the best trees were taken out, leaving genetically inferior ones to reseed the stand naturally. In many cases, the best practice is to do the reverse: take the worst trees out and leave the best to spread their superior genes around. Before the boom in woodburning stoves, it just didn't pay to remove the "worthless" trees. But now those trees are worth something, and a stand can be improved profitably. You get firewood now, and more timber in the years to come. Still, a firewood sale does not always pay off enough for you to hire a professional consultant to help. In this case, you can do a cruise yourself, deciding what to cut and when.

Managing for Timber

If your initial inventory has indicated that your woodland has saw timber, you will want to mark all the trees you plan to save for timber.

Cruising for Timber

Start ten to twenty feet from your property line or the edge of the woods. Mark a *crop tree,* a straight, tall tree that you want to save for lumber some day. Like the trees you marked in the sample plot, a crop tree should be free of signs of internal disease: swollen stem, seams or breaks in the bark, open wounds, or poorly healed branch stubs. After marking your first tree, pace off twenty feet parallel to the property line or the edge of the stand. Mark the closest crop tree, or one that might become a crop tree. If there are no trees within five to seven feet around you, take another couple of steps and try again.

Keep this up until you reach the end of the stand or the edge of the plot you want to work with. Turn at a right angle and pace off another twenty feet. Pick a crop tree, mark it, then turn again and go back, parallel to the first line. Continue this process, marking all the crop trees in your lot.

Selling Timber

A paid consulting forester will help you draw up a prospectus that tells the amount of timber you wish to sell. This private forester will also draw up a timber-sale contract and send it to at least six possible buyers. The buyer with the highest bid, or the most attractive logging plan, can then do the logging for you. This is called selling timber "on the stump." A good forester can mark the trees to go out (a government forester will occasionally do this for free), and can make sure that the logging techniques are environmentally sound.

Make your desires clear to the forester; if you want a beautiful old tree to stand, say so before the contracts get sent out. If you do not want an old stone wall to be torn down to make way for logging equipment, mention it before it is too late. In fact, most of these decisions should be made even before you hire a consultant. That way the government people can tell you whether a paid consultant is worth the money in the first place. Your timber sale should make at least enough money to pay costs for help and cover taxes on your profit. But if you are serious about producing timber on your land, local timber companies might give you free consulting work in exchange for first bids on the timber.

You can, if you want, sell wood directly to the sawmill. Some mills will come out and pick up logs that you stack by the roadside. You will get higher prices for cut wood, of course; but you need some expertise and some equipment. Some woodland owners have been known to rig their own log "deckers," which stack trees that have been stripped of their limbs. A few cannibalized truck parts and an old winch do the job. A good four-wheel-drive vehicle skids the logs to the site, with the aid of a logging chain and winch. Don't age the wood as you would firewood. Sawmills and kilns are expensive, so you will probably want to sell your logs green.

Trees that are not good enough for lumber can be used to make pulp, which goes into making paper, corrugated boxes, particle board, and many chemicals. Trees four to ten inches in diameter often make the best firewood; but some straight, tall trees of the right species can make fence posts and poles. Locust and ash make good heating wood, for example, but they are highly valued as fence posts that can stand without rotting for decades. Huge hardwood trees are sometimes cut individually for veneer and bring great prices. These logs are often shipped to Germany or Japan, where they are sliced into strips hundredths of an inch thick. Pines, particularly in the South, can be tapped for naval stores, and they make better pulp than firewood. But again, before you seriously consider selling any of your trees, get the advice of a good forester.

Maple Sugaring

If you have a number of maple trees growing on your land, you may be able to tap them for syrup and sugar — especially if you live in the North, from Wisconsin to Maine. Taps are put into sugar, black, silver, or red maples in the spring, just when the sap starts to flow. (Bigleaf maple and box elder can also be tapped.) The sap drips into buckets and is then boiled down into syrup.

Most maple sugaring is done part-time by farmers and small-woodland owners. Sugaring often complements tree harvesting for firewood and timber because the sweetest sap is produced by maples that have a little "elbow room," free from competition with other forest trees. The sweetest sap produces the best-tasting syrup, and more of it.

About thirty-three gallons of average-quality sap will produce a gallon of syrup, and the average tap produces between five and fifteen gallons of sap per season. A tree that is twenty-four to thirty inches in diameter can handle three taps without harm. In short, one very sizeable tree produces only about a gallon of syrup in a whole season. But the money is real good. Roadside stands in areas well traveled by tourists can command high prices.

Trees under ten inches in diameter should not be tapped at all. A tree with a ten-inch to seventeen-inch diameter can take a single tap, and a seventeen-inch to twenty-four-inch tree can take two taps. A tree larger than thirty inches in diameter can handle four taps. Taps should not be made along horizontal or vertical lines, or they might interrupt the flow of sap and weaken the tree. In following years, place new taps at least six inches away from unhealed tap holes; a hole takes from two to three years to close. Remove the spiles at the end of each sugaring season.

To boil the sap down, you need an evaporator, which is, essentially, a long series of pans heated from below. Sap flows from pan to pan, becoming more concentrated as vast amounts of water are boiled off as steam. You also need buckets, spiles or spouts, thermometers, filters, and sterile containers. Thermometers help you keep the temperature of the sap seven degrees above the boiling

Tapping a Sugar Maple

1. Drill a 7/16-inch hole, angled slightly upwards.
2. Tap the spile in gently.
3. Attach a clean, unrusted, gallon-size, covered bucket (or a milk jug) under the spile. You could attach plastic piping instead.
4. A 26-inch tree can handle three taps.

point of water during evaporation for syrup. Hotter than that, and you will make maple sugar.

You will probably want to do your sap boiling in an old shed, or some kind of shelter outside the home; that steam can get pretty messy. While you are figuring the cost of investment, count in the fuel it will take to run the evaporator. About half the cost of producing maple syrup is spent on fuel. If your woodlot is producing firewood, though, your problem is already solved. A small sugaring operation, producing just a few gallons of syrup per season, should pay for the original investment in several years, and still leave enough syrup for your Saturday morning pancakes.

Just a few trees can be tapped to make syrup in your kitchen as a hobby — if you are willing to risk peeling wallpaper from all the

The sugar maple is the best maple for tapping.

steam. You might even consider renting your maples to syrup gatherers. Some landowners charge twenty-five cents per tree per season. If the trees are tapped properly, they will remain healthy through many syrupy seasons. (For more details, see Bulletin A-51, Garden Way Publishing's *Making Maple Syrup* by veteran sugarman Noel Perrin.)

Recreational Uses For Land

You can make money from your woods without touching a single tree: by charging users for recreation. Many landowners are justifiably leery of visitors. Visitors have an annoying tendency to litter, leave gates open, tear down fences, shoot up signs, and rut trails with off-road vehicles. Worst of all, they sometimes get themselves hurt and hold the landowners liable — whether or not they were aware of the visitors' presence in their woods. But as the saying goes, when you've got a lemon, make lemonade. By selling user permits to responsible groups of people, you can establish direct contact with the people using your land.

Charging for Recreational Use

There are several ways to charge for recreational use. The most common — and profitable — is hunting and fishing fees. You do not own the wildlife on your land — the state does. But you do own access to that wildlife, and you can charge for it. Write your state fish and wildlife agency for local regulations.

With a small investment you can build a year-round campsite and charge for its use. A small cabin, piped-in water, outdoor toilet, and facilities for garbage disposal are a lot of trouble, but they may bring you more income than any amount of woodcutting you could do.

According to the Heritage Conservation and Recreation Service of the U.S. Department of the Interior, the fastest-growing sport in the nation is cross-country skiing. In parts of the Northeast, city folk are scrambling after uncrowded trails, which are harder and harder to find. If you have a trail or two with a total winding length of a few miles or more, you can lease trail rights. If your trail leads into your neighbor's property, you might consider banding together to rent the trails. Each landowner could be responsible for the trails on his or her own property.

Again, check your insurance coverge and local liability laws before you enter any agreements.

Managing for Wildlife

A great many woodland owners have just one other purpose for their woodlots beyond firewood. They use their forests for attracting wildlife. No matter how small your woodland is, you can manage it to attract and sustain a much larger number of animals than it presently does. Your state fish and wildlife agency can recommend ways to build nesting boxes and plant vegetation that attract wildlife. Some states sell game birds at a nominal cost for release on woodland. (A dozen quail make a great gift for the woodland owner who thinks he has everything!)

Your tree harvest can be worked around wildlife amenities. When you see an old, dead tree, think twice about cutting it up. The wood may be too rotten for burning, anyway; but the tree may be a wildlife heaven for the animals on your property. Woodpeckers and other birds feed on the insects the tree harbors. Raccoons, squirrels, owls, snakes, and other animals can use it for their dens. The U.S. Forest Service spends a great deal of money making sure that a minimum number of dead trees, called *snags*, remain standing after timber harvests over much of the national forest land.

The best kind of forest for wildlife has trees of all ages. It also has patches of seedlings or sprouts, called *openings*, interspersed with taller shade trees and cover. You can produce openings with small clearcuts. If you want to supplement the natural food with food of your own, make sure you *always* leave food. Many a deer has starved after the friendly couple that left oats in the winter did not spend the season in the country one year.

Help!

Conservationist Dorothy Behlen, in a column she used to have with *American Forests* magazine, described a program that the Tennessee Valley Authority designed in cooperation with the U.S. Forest Service. It is a computer program called WRAP, for Woodland Resource Analysis Program. For a small fee, the computer analyzes information collected by a forester who makes an on-site inspection; it then matches this information with the goals of the landowner, be they maintaining wildlife, scenery, fishing, hunting, timber, Christmas trees, or some combination of uses. The computer printout gives an inventory of the land, a list of the management practices needed to achieve the landowner's goals, and a year-by-year plan for each acre of land.

Unfortunately, WRAP is available only in Alabama and Kentucky. But it may someday be available for nationwide use. In the meantime, you can get some free help to devise a program of your own and come up with an inventory, list of management practices, and step-by-step plan. Public foresters do not have a whole lot of time to spend with each landowner, and it may take a while to get one to come out; but the free advice can help steer you to a decision on whether to hire a consulting forester.

Free Help

For managing the land, a lot of free government advice is out there for the taking. The following agencies offer forestry help.

State foresters. Many states have a forester in every county. The department or division of forestry is often contained within the department of natural resources or environmental resources, or the department of agriculture.

AMERICAN FORESTRY ASSOCIATION

Membership in the American Forestry Association, a conservation organization that concerns itself mainly with forest resources, will get you a full-color monthly magazine, *American Forests*. The magazine usually has at least an article a month of interest to the woodland owner, and it often runs articles with facts on trees and forests. In addition, the association has a free question-answering service for its members. Write American Forestry Association Membership, 1319 18th Street NW, Washington, DC 20036.

State woodland-owners' associations. Your state forester can help steer you toward landowners who have banded together to share advice, buy equipment in bulk, and lobby local and state governments. Such organizations are springing up all over the country, particularly in the Northeast and parts of the South. Some of them are affiliated with the state forestry association, whose address the state forester can give also.

U.S. Forest Service. This federal agency works with states in the Rural Forestry Assistance Program. It gives good advice through land-grant university extension foresters. And it doles out financial assistance to woodlot owners through the Forestry Incentives Program and the Agricultural Conservation Program. The Forestry Incentives Program allows the federal government to share up to 75 percent of the cost of planting trees and improving a small owner's forest stand. For the nearest forest service office, look in the phone book under the U.S. Department of Agriculture.

U.S. Soil Conservation Service. The Soil Conservation Service can give you advice on soil conditions of your land. Like the Forest Service, it is within the U.S. Department of Agriculture.

Professional consultants. Timber companies often provide professional help to landowners, with just one string attached: the company gets first refusal on the sale of the timber. Write to the American Forest Institute for its *Directory of Forest Industries Providing Forest Management Services for Private Landowners.* Their address is 1619 Massachusetts Ave. NW, Washington, DC 20036. If you have ten or more acres of woodland, you might join the forest industry's tree farm program, run by the American Forest Institute. It costs nothing to join, and you get the *Tree Farm News,* a bulletin that often gives down-to-earth advice to the woodland owner and lets you know what other folks are doing with their woods. Write Richard Lewis at the above address.

If you decide to hire a consulting forester to come out and give some management advice, be prepared to pay twenty-five to forty dollars per hour (but fees vary greatly). For a nationwide list of consultants, write for the *Directory of Consulting Foresters* from the Society of American Foresters, 5400 Grosvenor Lane, Washington, DC 20014. The Association of Consulting Foresters is an organization that sets professional standards for its members. It puts out a *Membership Specialization Directory,* which is more detailed than the general one just mentioned. Write Edward Stuart Jr., Association of Consulting Foresters, Box 369, Yorktown, VA 23690, for a copy.

Taxes

If you have not done so already, look into your tax situation. Many a conscientious woodland owner has been forced into environmentally unsound practices — or even into sale of the land — because of poor tax planning. They say that nothing is as inevitable as death and taxes, but modern medicine can prolong life well beyond ancient expectations. It is not the same for taxes; no amount of political wizardry has yet managed to make a landowner feel chipper come April.

If you market firewood or any other product of your land, you will have to pay federal tax, and probably state tax as well, on the income. But profits on sales of your timber can be classified as "capital gains," and you can avoid paying more than half the tax on it. If the woodcutting was done to improve the woodlot, however, added value of the forest could increase your property tax — except in areas where innovative laws have alleviated this problem. And do not forget estate taxes, which could force your heirs to butcher the woods or sell some of the land to get the government off their backs.

In many cases, you can write off, or delay paying, taxes on some of your woodlot expenses, such as tools and reforestation. A few months before President Jimmy Carter left the White House, he signed a bill that includes a reforestation incentive. This law gives a 10 percent tax credit for reforestation expenses — site preparation, labor and tools, and depreciation of equipment used in planting or seeding. For example, if you spend $2,000 to replant cut-over woodland on your property, you will be allowed to subtract $200 from the taxes you would normally pay that year. In addition, a seven-year amortization on the first $10,000 of capitalized reforestation expenditures each year allows further tax relief. What all this means is that you can deduct a certain amount when you state your adjusted gross income each year for seven years on your tax form.

Most important to a woodland owner, particularly an elderly one, is estate planning. Without good planning for the future, all your woodland work could go down the drain. But, by giving parts of your land to heirs in parcels every year, you could end up saving them thousands of dollars in taxes later.

To help you with all these problems, find a lawyer familiar with timber taxation. Write your state forester or your state bar association for referrals. Some timber-tax lawyers even specialize down to estate taxes. Figure out what kind of advice you want in advance. To help you understand some of the tax jargon, write for information on timber taxes to American Forestry Association, 1319 18th St. NW, Washington, DC 20036.

The final solution to paying property tax is to avoid it, by getting the government to buy or trade tax relief in exchange for rights to your property. If your woodlot is important to a city's watershed, or to an area's aesthetic value, or if it has special qualities like wildlife or rare plants, you might be able to get the state or local government to agree to a *conservation easement*. In return for your promise not to develop the land, the government relieves you of the tax burden. In many cases, landowners are allowed to cut their timber and to exercise other good management practices. Write the Nature Conservancy, 1800 North Kent Street, Arlington, VA 22209, for more information on easements.